T0044872

Father Sky and Mother Earth

by Oodgeroo

WILEY

For all children
and especially my grandchildren,
Joshua and Ché Walker

ONCE UPON A WORLD

Father Sky and Mother Earth had four children: Sun, Moon, Sea and Rock. They never again lived together after they had their children.

Sun and Moon created servants for themselves. They named them Clouds, Winds, Rain, Stars and Storms. Sun and Moon and their Servants lived with Father Sky.

Rock created servants too. He named them Tree, Birds, Animals, Reptiles and Insects. Rock and his servants lived with Mother Earth.

Sea's servants were Oceans, Tides, Currents and Gales. Sea and his servants lived between Father Sky and Mother Earth.

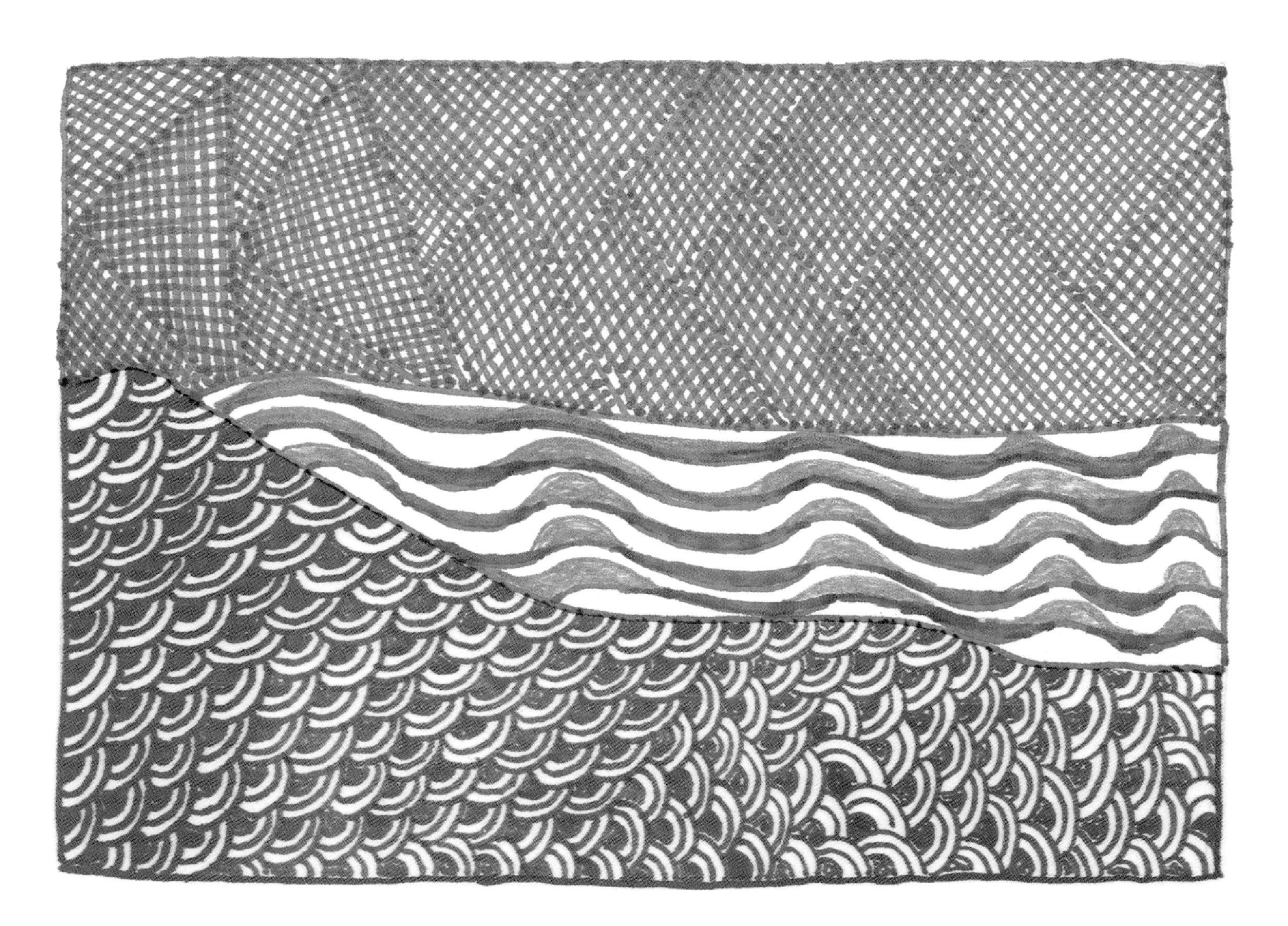

Father Sky watched over Mother Earth. When she grew thirsty, he asked Moon and Sun to send her water. They sent Rain to help. Rain created Rivers, Creeks and Lakes and filled them with water.

Sun sent his warmth to Mother Earth. Rock, Trees, Birds, Animals, Reptiles and Insects all loved Sun.

In the early morning and early evening, Sun would light up Father Sky with his many colours. Sun called his early morning colours Sunrise and he called his early evening colours Sunset.

Moon and Stars lit up Father Sky at night after Sun had gone to sleep.

Sea's servant Gale looked after Father Sky and Mother Earth by keeping everything clean. Gale knew he needed help, so he created Cyclone and Tornado to help him.

Sea's other servants, Oceans, Tides and Currents, would be blown and tossed about, and would go wherever Gale, Cyclone and Tornado blew them.

Gale, Cyclone and Tornado sometimes blew so hard that they knocked over Rock, his servant Tree and other things that got in their way.

Rock's other servants, Birds, Animals, Reptiles and Insects, would hide until Gale, Cyclone and Tornado had finished their cleaning.

Birds lived in Tree's branches and some Animals would climb up Tree's trunk and rest there.

Insects lived under Tree's bark or on Tree's trunk and branches.

Gum Tree let Koala live in his branches and Koala fed on Gum Tree's leaves.

Other Animals and Insects liked to burrow under Rock or sit on him and warm themselves.

And Platypus lived with Frog and Tortoise in Rivers and Creeks and Lakes.

Rock created Mountains and Hills to protect his servants from the cold winds of Gale, Cyclone and Tornado.

And Tree created Plants and Grass and Flowers.

And Animals, Reptiles and Insects created more Animals, Reptiles and Insects.

And so on and so on and so on ...

And they were all very happy creating and balancing and loving and living and helping one another.

One day, when Trees, Plants, Grass and Flowers were busy looking green and bright, Bee, whom the Insects had created, came along.

Bee explained that if he took pollen from one Flower to another Flower this would help Trees, Plants, Grass and Flowers to create more Trees, Plants, Grass and Flowers.

Besides, Bee told them, he could also create honey in his home in a hollowed-out Tree, and this would help feed the Birds, Animals, Reptiles and Insects.

Trees, Plants, Grass and Flowers welcomed Bee and told him to come as often as he liked.

And in this way they all helped one another.

Sea created more servants, and they lived with Sea. They were Fish, Shellfish, Starfish, Seaweed and Crab.

And Octopus lived with Sea. Octopus dug a big hole in Sea's bed and scattered rocks and shells all around it. Octopus lived in this hole.

And Mangrove Tree lived in the mud swamps close to Mother Earth in Sea's salt water. He sheltered and fed Fish, Shellfish and Crab. Sea Birds nested in Mangrove's branches. Mangrove grew little spikes in the mud to help him live and breathe.

And Whales played in Ocean’s vast waters.

And so it was a happy time of Creation, and those who came after Sun, Moon, Sea and Rock kept creating.

And Father Sky and Mother Earth were very proud and happy for them all ...

AND SO
THIS WAS
THE BEGINNING

Father Sky and Mother Earth, their children Moon, Sun, Rock and Sea, and their children's servants lived in peace and happiness and there was beauty everywhere.

Until ...

Until ...

One time a strange Animal came among them. This Animal was not like any Animal they had ever seen and this Animal was called Human.

And Human Animal cut down many, many Trees to build houses. Koala, Birds, Reptiles and other Animals became afraid of Human Animal and fled for their lives.

And Bee flew away because Human Animal spilled Bee's honey when Human Animal cut down Bee's Trees.

And Platypus, Frog and Tortoise couldn't enjoy their swims any more because Human Animal threw all his rubbish into Rivers, Creeks and Lakes.

And Human Animal made boxes with motors in them and they made a lot of noise. The motors blew out smoke that choked the Animals, Trees, Birds, Reptiles and Insects.

Human Animal's boxes ran all over Mother Earth. And some of Human Animal's boxes flew high in Father Sky – higher even than any of the Birds.

And Human Animal's boxes filled Father Sky with smoke.

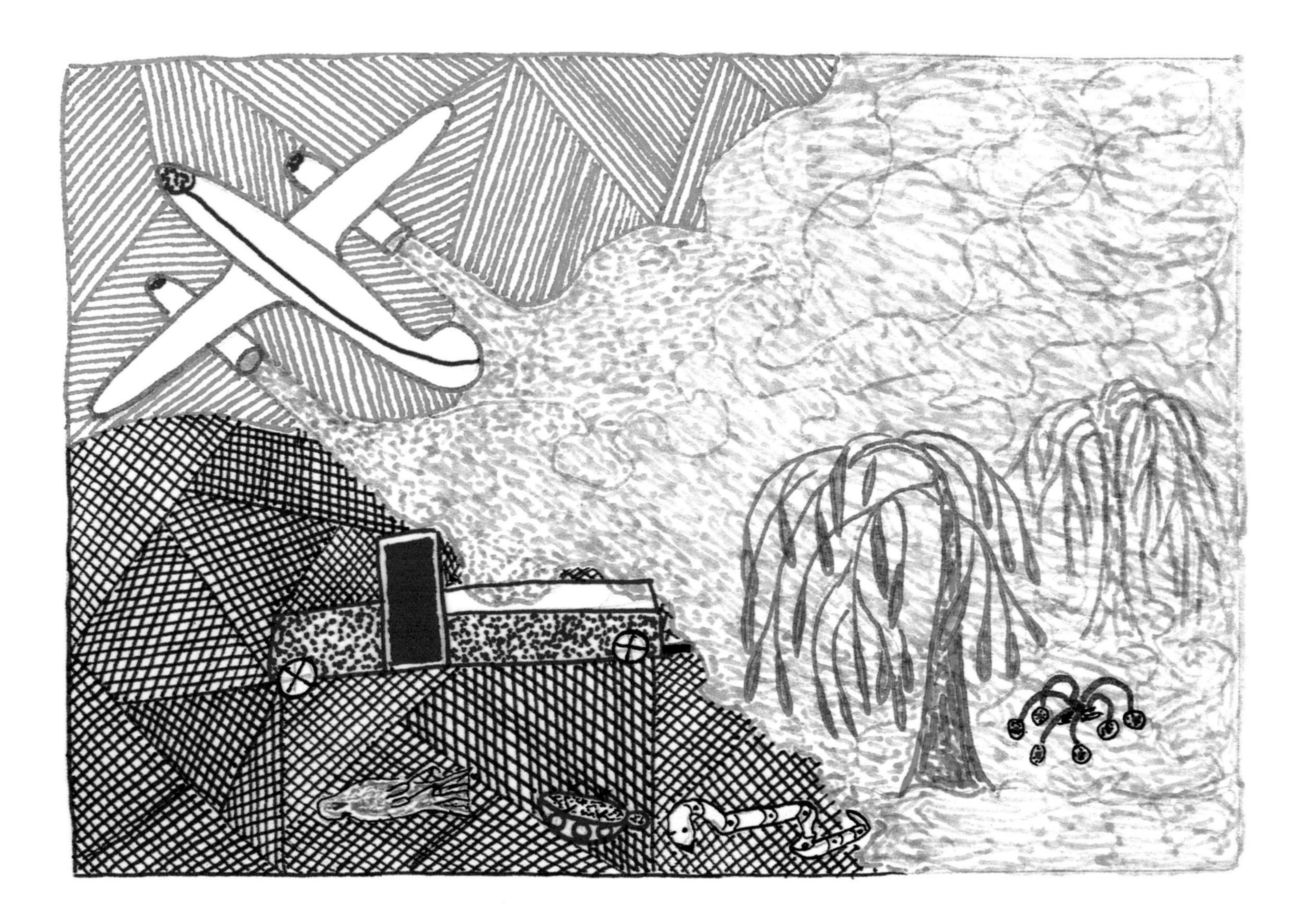

Some of Human Animal's boxes ran all over the Sea and spilled oil and smoke everywhere.

And Human Animal's boxes chased Whale all over Ocean because Human Animal wanted to steal Whale's oil.

And Whale was so busy keeping out of Human Animal's way he no longer played in Ocean's vast waters. Whale just swam for his life.

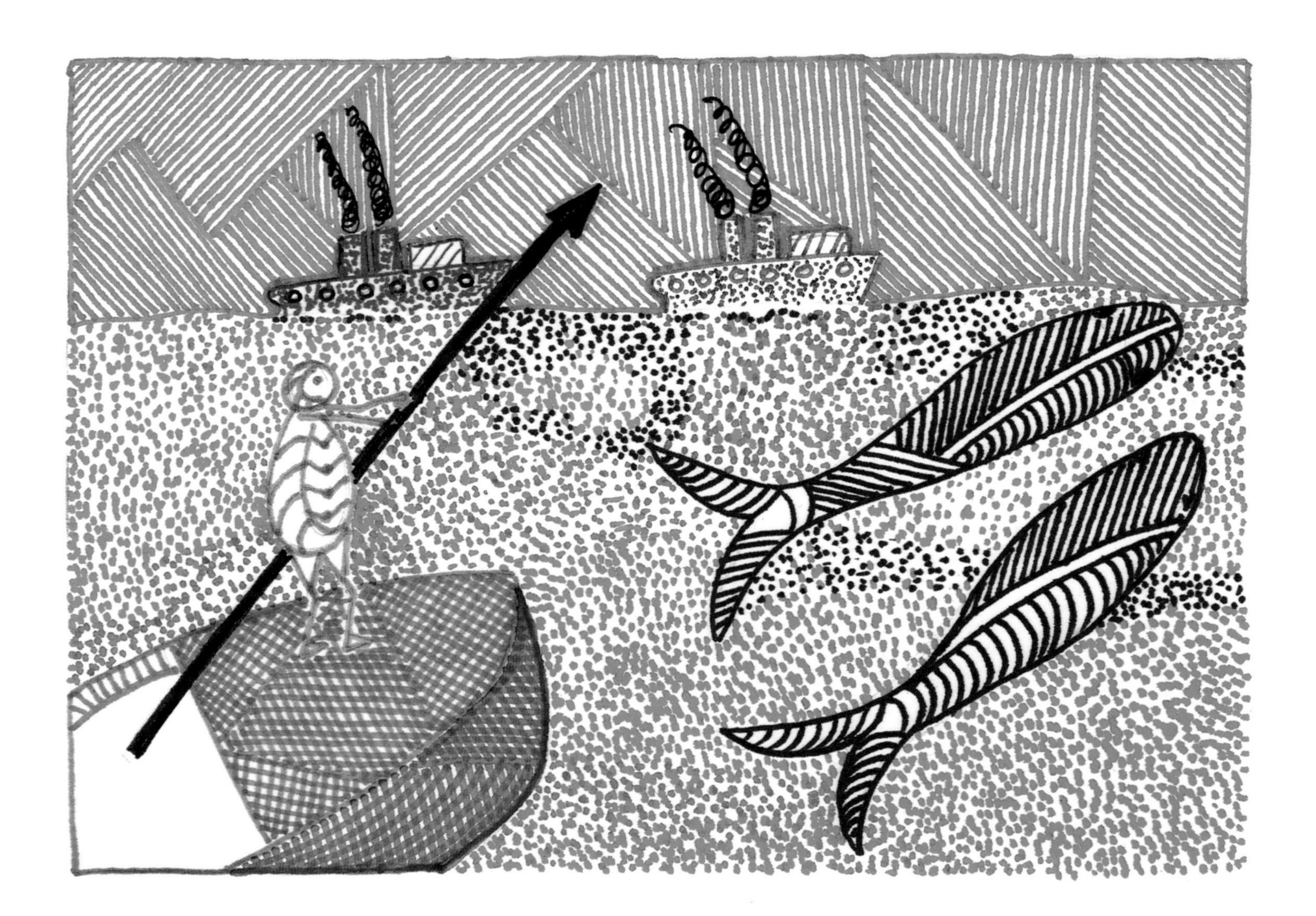

Mangrove's leaves turned white and curled up and fell off because Human Animal covered Mangrove's spikes with strange sand when Human Animal filled in Mangrove's swamps.

And Seaweed didn't grow very well because the strange sand choked Seaweed.

And Fish had to go far away to find his food. So he swam to an island way out in Sea, where other Mangrove Trees grow.

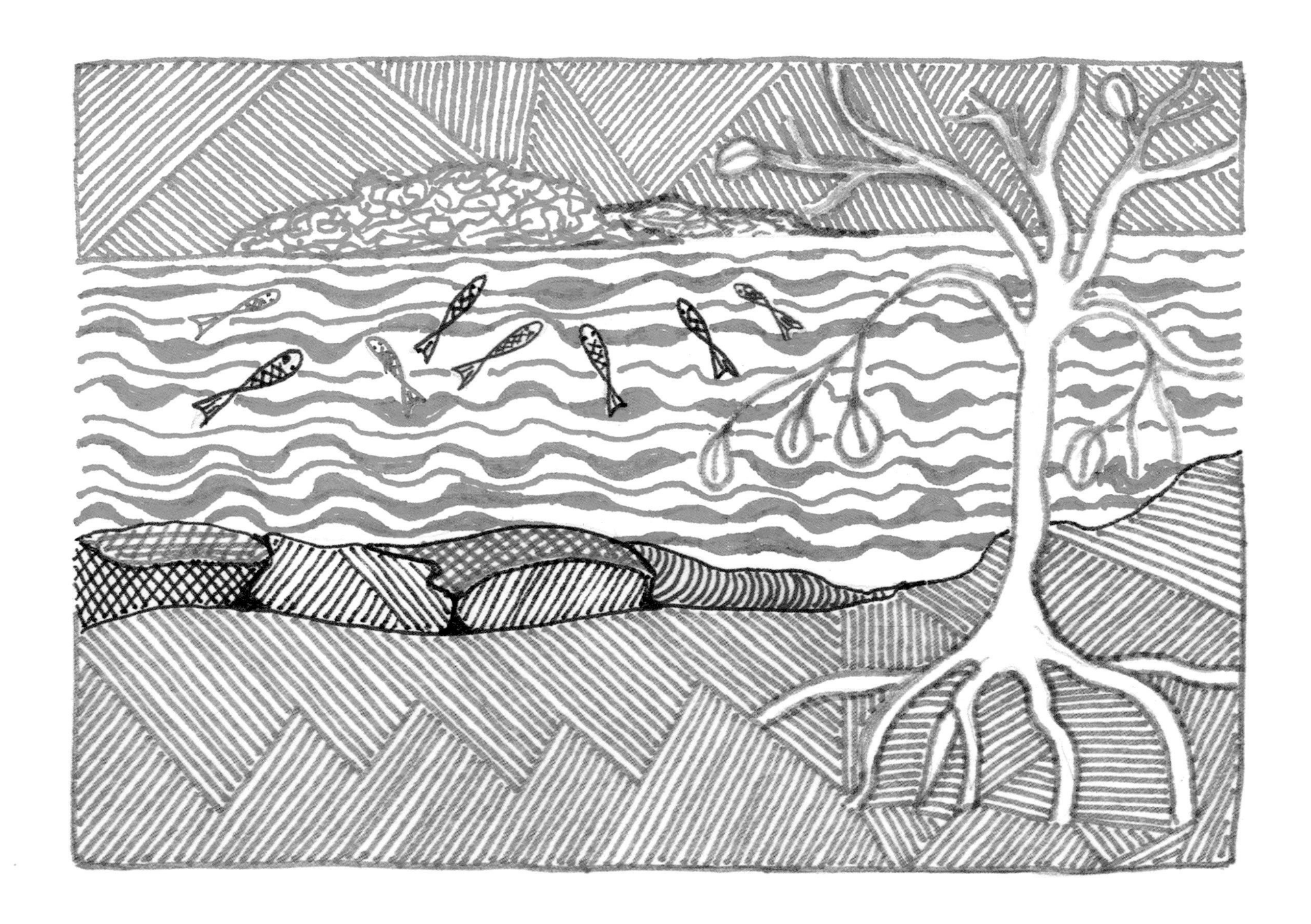

And Crabs had to leave their home under Mangrove. Some Crabs went to live with Coral.

And Starfish crept back into the deep, deep part of Sea. Sea hid Starfish to protect him and Starfish lived with Sea Cucumber.

And Octopus couldn't live in the hole he dug because the strange sand kept filling it in. So Octopus had to leave and he went to live with Sea Cucumber and Starfish in the deepest part of Sea.

Then, one day, something happened.

Some Human Animals looked around and saw all the damage Human Animals had done. The mess made them very worried and sad. So they decided something had to be done about it.

All the worried Human Animals got together and had a big talk-talk on top of a hill.

The worried Human Animals decided to clean up all the mess that Human Animals had made.

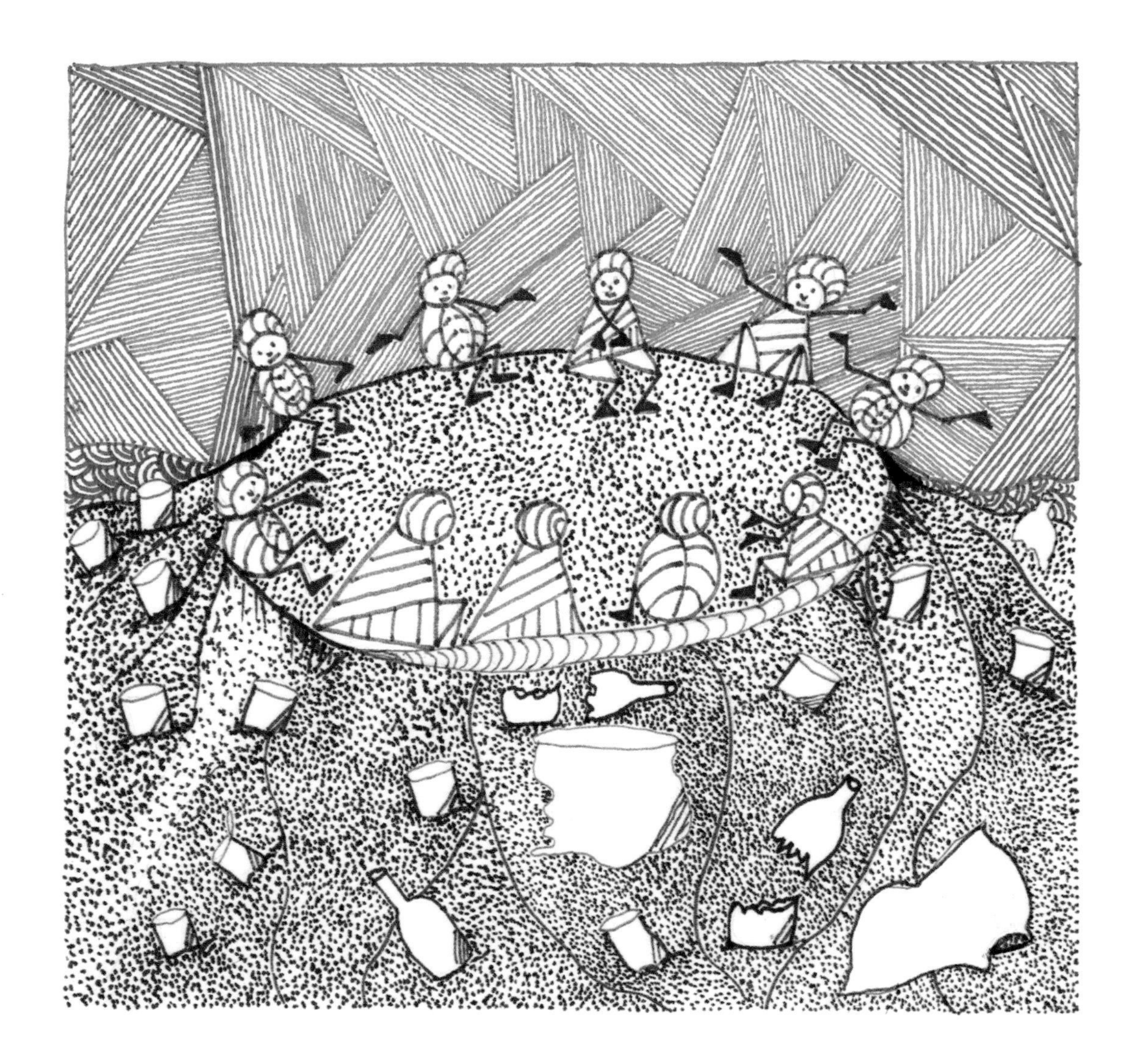

The first thing the worried Human Animals did was put up warning signs. The signs read

BEWARE!
Human Animal is the
most dangerous Animal of all

And the worried Human Animals put up many warning signs all over Mother Earth.

HUMAN ANIMAL DESTROYS
HUMAN ANIMAL IS THOUGHTLESS
HUMAN ANIMAL POLLUTES
HUMAN ANIMAL RUBBISHES
BEWARE HUMAN ANIMAL IS THE MOST DANGEROUS OF ALL
HUMAN ANIMAL MAIMS

And the worried Human Animals worked hard, too. They gathered up all the rubbish they found in the Creeks and Rivers and Lakes and wherever else they saw rubbish on Mother Earth. Rubbish was everywhere and it kept the worried Human Animals very, very busy.

And the worried Human Animals all wore badges that read

ARE YOU
A WORRIED HUMAN ANIMAL?
I AM!

ARE YOU A WORRIED HUMAN ANIMAL? I AM
ARE YOU A WORRIED HUMAN ANIMAL? I AM

And perhaps, when many more Human Animals join the worried Human Animals and work with them ...

Soon, very soon, perhaps ...

Father Sky, Mother Earth, their children and their children's servants will once again be bright and clean, and happy and beautiful and peaceful, just as they were in the beginning.

And perhaps Human Animal will be able to live happily and peacefully with Father Sky and Mother Earth and their children.

And perhaps there will be ...

A NEW
BEGINNING

Fourth edition published in 2021 by
John Wiley & Sons Australia, Ltd
42 McDougall St, Milton Qld 4064
Office also in Melbourne

Typeset in GaramondPremrPro 20/26pt

First edition 1981
Limp edition 1985
Third edition 2008

ISBN: 978-0-730-39113-5

A catalogue record for this book is available from the National Library of Australia

Cover art by Oodgeroo

Printed in Singapore by Markono Print Media Pte Ltd

10 9 8 7 6 5 4 3 2 1